Digging for Profit

Secrets to Success in the Construction Industry

Jeff Spencer

with Jerry Work

I0493040

Digging for Profit: Secrets to Success in the Construction Industry

Table of Contents

Digging for Profit: Secrets to Success in the Construction Industry

Introduction

Construction is a tough business. It can make you rich. And it can drive you into the ground. I should know. I've been in the construction industry for nearly a quarter century. I've been the guy sitting in the office bidding jobs. I've been the guy operating the backhoe. Hell, I've even been the guy shoveling gravel. I've made a lot of money. And I've gone bankrupt.

But I dug my way back out. I wouldn't trade any of that experience because it has made me who I am today (well, maybe I would trade the bankrupt part). Today, I am a professional construction estimator and co-founder of ProfitDig LLC - an online application for easy construction estimating and job costing.

This book is the culmination of my nearly twenty-five years' worth of knowledge in the construction industry.

You can be successful in your own construction business, if you do things the right way.

First off, you need to be really good at whatever it is you do. Forget about starting your own business until you have mastered your craft.

Second, you need to be organized. You need to have a system for creating job bids that accurately reflect what it will cost you to complete the work (with some profit on top). And you must have a way to track all of your costs against the items in your bid.

Third, you need to be professional. If you want to grow your business, people need to respect you and like you. You need to deliver high quality results on time and on budget. And if on time, on budget isn't going to happen, you need to know how to deal with it.

These are the things we will talk about in this book. So let's get to work!

Chapter 1 – How to Get Started in the Construction Industry

So you've decided to seek fame and fortune by starting your own construction company. As with most things, it's easier said than done. Before you even THINK about it, you better make sure you are very skilled at what it is you do. Don't even think about trying this until you have mastered your particular construction niche. So assuming you have the technical skills, how do you go about actually starting your own construction business?

To start with, you've got to really put some thought into this process. Don't just jump into it without being extremely prepared. The first thing I would do is build a portfolio of my work. This is going to be necessary when talking to banks, and it may even be something that larger contractors (the guys who might hire you as a subcontractor) may want to see as well.

By portfolio, I mean a document with examples of the work that you have done for your present employer. It should also have a description of what you're looking to do, the type of income you are seeking, and the size of the business that you want to start off at. Financial stuff, in other words.

There are a lot of factors that come into play now. You are about to make the plunge. This is reality. Now you really need to start looking at things like:

- Your resources. Do you have any money put away? How much are you going to need to borrow?

- Insurance requirements. This is extremely important. It's a law that you have to have workman's comp insurance. You need to be getting quotes on this.

- The cost of equipment. You need to get prices on whatever equipment you think you need.

- Are you going to rent or purchase your equipment? Will the bank work with you? Will the equipment people work with you?

- How much manpower do you need? Are you going to do most of the work yourself? With a handful of people as a starting crew? It's probably a good idea to start small. But you've still got to have transportation to get back and forth to work and to carry your tools. If you're going to have multiple crews, then you're going to have even more expenses to give those crews the resources they need.

- Cash flow. You're used to a certain amount of money coming in every week, just like the people you're looking to hire to run your crews, or be your crews. They're used to being paid a certain amount of money every week. So you've got to make sure you've got financing available, because it could take you four, five, six months to bid work, pick up a job, and actually start it. You've got to sustain yourself in the meantime. I would not even think about hiring a crew until I had my first job secured.

How do I get an idea of how much money I need to start my business?

From early on, you've got to do your homework and have a specific amount of money in mind. You need to have a vision for what you want your company to be and what size you want to be before you ever go talk to your potential clients.

Let's say you're looking to start off small - half a million dollars a year of work. Combined with the type of work you are going to do, that will give you a rough idea of how many men you're going to need. Let's take an excavation company, for example. In excavation, it depends on what size projects you're going to look at. If you're going to do a half million dollars a year, you're going to do some small site jobs. A three to four man crew and three to five pieces of equipment is going to be sufficient to do the kind of work you want to do.

If you go with rental equipment, it's pretty simple to figure out what your budget is going to be as far as what your equipment is going to cost you just to rent it. Same thing if you're going to purchase equipment. You talk to your equipment salesman and find out what it is going to cost. Tell him, "I need this, this, and this number of pieces. What's it going to cost me?" You'll get an idea - a rough idea, depending on your credit - of what your monthly payment is going to be.

Then you've got to know how much fuel each piece of machinery is going to burn. The way I look at it is long-term. Let's say I've got a piece of equipment sitting out here that's going to burn 75 gallons of fuel a day. Well, current fuel prices may be $2.50 a gallon, but I'm going to add an extra fifteen or twenty percent to that just to make sure I'm covered if fuel prices go up. If fuel prices go down, then that's just putting more money in my pocket.

In addition to machinery, there are hand tools that you're going to need. You will need pickup trucks to get you to and from the job. There may be safety equipment you need.

Say you're into excavation work, and you're doing trench excavation. You may have to have a trench box. You may be working in the street, requiring you to rent street plates.

You need to think through every part of your business. There are a lot of variables that you've got to figure in your original budget to make sure you have enough money to operate. You've got to do your homework, and figure out what your equipment is going to cost, fuel to run it, employee wages, taxes and insurance, and any other overhead or expense necessary to conduct business.

Renting versus buying heavy equipment - what's the best way to go?

In my personal opinion, I prefer to buy equipment. Many equipment companies have lease to purchase options. If you're not financially stable enough with the bank to make that purchase yet, you can still lease a piece of equipment. The lease is going to cost you more money per month. But if you can do a lease to purchase deal then the equipment rental company will let 100% of what you pay on that lease go towards the purchase of it like a down payment. I would recommend that kind of option, or if you're able to just to purchase it outright, that's the way to go because you're going to save at least 40% on your monthly payments versus leasing it.

We will talk more about this in more detail in a later chapter.

How Do I Find a Crew?

It's tricky putting your first crew together. A lot of times it's really hard to find help, especially in an area where there is a lot of construction going on. There is a lot more competition for men in that case. If you've been in the construction industry for a while, you may have a group of men that will follow you wherever you go. Whenever you leave your job, they may come with you. Or through your resources you could pick up a man or two here and there to fill in for what you need.

That kind-of goes back to the whole idea of doing the best work you can possibly do at your present job, which can get you in the door with some possible customers as well as getting your crew together. I can't emphasize enough that you need to be a high value employee for your present employer, and build respect both with the men you work with as well as your customers.

Being the type of person who treats his employees with respect goes a long way. Construction is hard. A lot of supervisors are on tight budgets with tight deadlines, and they can be pretty harsh sometimes. Sometimes you have to be, but you can also gain the respect of the people that work under you in the way that you go about getting your point across. If you ever expect any of those guys to follow you when you start your own company, you need to treat them well.

Chapter 2 – How to Get Your First Customer

"I see we're going up against the Big Guys."

So you've started your own construction company. Great! But now you need a customer. How do you go about getting that first customer?

To start with, you would want to go around and speak with companies that you're already associated with, that you've done work for in the past, or that the company you currently work for has worked for in the past. Talk to some upper management type people, like project managers, estimators, and owners. Tell them what your ideas are, what kind of company you're looking to start and just ask for help. Tell them you want to sub some work from them. A new company starting out fresh is going to have a hard time just going out and picking up work without having someone on your side to be able to help you out.

In other words, before you just go and quit your job and start a business, you should try to leverage contacts you can make through the job that you currently have.

And there's nothing wrong with that. You're not trying to take work away from the company that you're working for. You're just asking for an opportunity to bid on some jobs. Whoever you talk to, they're not going to just come out and hand you work. You may have a good relationship with somebody who might do that, but for the most part, you're just asking for the opportunity to bid on their work. If you have a competitive bid and can negotiate with them, then you have as much right to do that as the next guy does.

To start with, you need to make sure that you do really good work and that you can represent yourself well to people you may want to try to get work from. Say you're working in the field as a crew leader or possibly a superintendent and you don't want to start your business for a while. Eyes are always on you. You want to do quality work even when you're working for someone else. It's very important because out of respect, you'd want someone working for your new company to give you the same as what you expect.

The people that you will eventually go talk to and ask for help will be watching you, especially if you're working on one of their projects. Once you contact them for help, they're really going to be watching you. Is he really as qualified as he says he is? Does he really know everything that he says he does?

Again, be really good at the job you do and then make contacts through your job and present yourself professionally when asking for that first shot.

There are always situations where you're on a project – whether you're a project manager or a superintendent or a foreman – when your supervisors are telling you to do something one way but you may see an easier way to do it - a way to save the company money - but you might be getting overlooked. Little things like that are just good rules of thumb to store in the back of your mind (or on paper). Think to yourself, "If I ever have my own business, this is the way I want to do things."

You want to try to optimize every situation, to get the most out of it for you or whoever you're working for, to try to be as profitable as possible.

One thing we have not mentioned is advertising. In my opinion, I would avoid general purpose mass market advertising – or "branding." You are unlikely to have the kind of budget you would need to effectively advertise your business on a large scale. I suggest you start out by leveraging your existing contacts and work history to generate bid opportunities, leads and referrals. If you do any kind of advertising, make it as specific as possible. One alternative to mass market advertising is direct marketing, such as sending a sales letter or brochure to a specific list of potential clients.

Chapter 3 – How to Bid Your First Job

So far we have talked about how to start your own construction company, the resources you'll need, and how to go about finding your first customer. Now let's discuss how to actually go about bidding on your first job.

This is where it gets tricky. You may have been in the field. You may have done the work, and been good at doing the work, but knowing what it actually costs the company you work for to do that work is a totally different thing.

You know you're going to have overtime. You know that you're going to have to have insurance. You know you're going to have to have certain pieces of equipment like vehicles and trucks. You're going to have things like office supplies, office help, etc. that are not going to be associated with particular jobs. Working through these items is how you come to grasp what your overhead is going to be.

It's hard to figure out your overhead starting out because you don't have anything to compare it to. Once you've been in business for a year, you can look back at what your cost was related to your profit, and it's simple math. Once you have some data to work with, you can go back and figure out what doesn't get charged to jobs so you know what your actual overhead percentage is.

Most small companies end up with roughly 5 to 7% overhead. It's going to vary in different states and different areas of the market you're in, but that should be a pretty good rule of thumb. 5 to 7% should cover your overhead.

So you need to know your overhead, and you need to know the value of your work. Once you've made it known to contractors that you're getting ready to start bidding on work, you don't want to turn around and look like you don't know what you're doing. Your potential sources of work will start second guessing you, and they may very well decide they don't want to hire you.

You've got to know how long it's going to take you to do the work that you're about to bid on. You've got to know how many pieces of equipment you need. You've got to know how many employees you're going to have to have, and what the time frame is.

All of this comes from experience. So again, don't make the leap until you have gained the proper experience on the jobsite to know how to bid a job.

Every job has a deadline. To make money, you want to beat that deadline. So If you've got 120 days to do this job, that's where your time is going to start. You've got to figure out how many hours you're going to work per day. Whether it be 8 hours, 10 hours, 12 hours, you've got to know that up front. Then you can divide that by the amount of yardage you've got to move, or the amount of pipe you've got to put in, and so on, to get to the point of knowing what it's going to cost.

What if it takes you longer to do a job than you have accounted for in your bid?

To start off, engineering is not always fail proof. They make mistakes like everybody else does. Sometimes things don't work and have to be changed, and at that point, you get to rebid a portion of your project to add more money into your current contract.

But if nothing changes and you do not complete the project in the amount of time that was allowed or that you bid on, then it's a life lesson. You've lost money. Hopefully you haven't lost enough that this is your last job as well as your first. The goal is, whatever your deadline, to try to beat that. You want to come in under budget.

That's another thing that you're going to have to figure out - how you're going to track costs - and to come up with that budget based on your expected costs. Every day that you finish ahead of schedule is that much under budget you're going to be and that much more profit you're going to make off of that project.

Generally, if you have a deadline on a project, you need to make sure that you're going to be under that deadline. Most of the time your engineers or your owners or your municipalities - depending on whom you're working for – will have a liquidated damages clause in their contract with you. That will state that if you do not finish by a certain date, every day after that, you've got to have substantial completion within a certain number of days. If you go over that, you will have to pay a certain amount of money every day. It may be $500 a day. It may be $200 a day. It could even be $1,000 a day, but it's in your contract that you'll have to pay that much for every day you go over. Not only are you losing money because you underestimated a project, but now you're losing money because your customer is charging you for every day you run long.

So at this point, you've quit your job, you've started your own construction business and bid on a job, and you're doing work. How do you keep track of all this - your bids, your costs, etc.?

At first, tracking all of your information is going to really be tough, especially if you are physically doing the work yourself. A lot of people who start off from scratch have to work in the field for a few years to get established enough to hire crews to actually do the field work for them. There are numerous ways to track your expenses. You can make up your own spreadsheet format and try to keep up with daily time and writing down your costs. The problem is if you're working 10 or 12 hour days (which is not unusual), it's tough to come home at night and log all this stuff so you have an accurate account of what you actually spent that day.

There are a lot of different avenues as far as software you can buy for estimating and job costing and such. and they can get pretty pricey. In my biased opinion, I have to recommend ProfitDig because it's both tools in one – bidding and costing. That's the best part about it. And it's a non-contract, low monthly cost application. Companies in this industry have traditionally had to rely on either manual spreadsheets to track costs or very expensive desktop software. You have the advantage of starting your business in an age where there are much better, more convenient and lower cost solutions.

Using a tool like ProfitDig, while you're creating your estimate to turn in to your client to bid a potential job, you are creating a budget at the same time. The application takes your raw expected costs and logs it at the same time that you're entering it to give you a working budget. Then when you actually enter your time and costs on a daily basis it will track it against your budget. Materials bought and any other kind of costs that are associated with that project will be recorded in real time.

What goes into a bid?

To create a bid, you have to know your bid items – especially if you are going to use a program like ProfitDig or something similar to create the bid.

Bid items are items that you ordinarily bid on in your line of work, whether it be structural work, excavation work, underground utilities, concrete - whatever. Everybody is going to have their own set of bid items that they normally bid on depending on their field.

Let's take excavation as an example. With excavation, your standard bid items will start off with mobilization. It costs you money to move into a job. You've got X pieces of equipment you've got to move in and fuel associated with that. You have labor costs that you have to associate with that because those trucks don't load equipment and run themselves up and down the road. You've got to have people to actually do the work.

If you have big equipment, you have permits you've got to buy to move that oversize equipment up and down the road. It takes special insurance to move that oversize equipment up and down the road. Sometimes it takes escorts. There's a lot of stuff to take into account. And that's just one bid item getting started. You've got a lot of stuff to think about as far as what it's actually going to cost you.

You may have a hundred bid items in your categories that you normally bid on.

When I'm estimating a project, the very first thing I look at is mobilization. It's not just to move in. A lot of times jobs share phases. You may come into this one job and it's got three phases. Not only do you have to move in, you've got to move out. Then you have to move into phase 2. Then you've got to move out again. Then you've got to move back into phase 3. Then if you don't have another job to go to you've got to move the equipment back to your yard or somewhere else, and that's going to cost you money.

Also in mobilization, if you've got three trucks running down the road every day with people riding in them to get to and from the job, then you've got fuel associated with that. You've got to be able to figure what it's going to cost you to get those pickups up and down the road. All that stuff goes into your mobilization price.

When setting up a bid, or an estimate, bid items are not really the top level items. There is a level above that. Those top level items are called "headers."

Say it's erosion control. That's going to be one of the first things you bid on. You're going to have to put up a silt fence on the project. You're going to have erosion hills. You may have inlet protection. A lot of times you have a preliminary and a secondary erosion control plan, depending on how your engineer has designed it. You also have to do erosion control monitoring. So you may have ten bid items under your erosion control category. That's one header.

So you put in those bid items for that header, and then you go to your next header, which would be grading and excavating. Then you enter your bid items for that one. Then your next one may be storm drain. Same thing, keep on going - water, sewer, whatever that project may consist of.

How much or how little detail you put into your bid will depend on the client. If you're bidding DOT work, you'll have to show your work, your line items, values and what each one costs. If you're bidding to an owner or specific entity that's not associated with a government, then they're not necessarily going to require you to show that level of detail. You don't necessarily have to show your extensions. You can just show subtotals for each category, such as grading, erosion control, or storm drainage.

Just showing subtotals gives you a little more flexibility if the client wants to take something out of your contract. If you haven't broken everything out in your bid, then you can still keep some of that profit in your bid even if you are cutting an item out at the client's request. You don't have to give it all back because they don't know exactly how much you had for that particular item.

A bid needs to include manpower, equipment, materials – everything. If you're bidding grading, most of your expense is just going to be labor, fuel and equipment. But when you get into storm drainage, water, sanitary sewer, electrical, fire lines, concrete work and structural work, your materials add up pretty quickly.

There is going to be a fairly standard set of bit items that most of your bids are going to have, depending on your particular specialty. So your job estimating will go a lot quicker if have some kind of system where you're not having to write (or type) those all out from scratch every time. This is another area where a solution like ProfitDig can make your life a lot easier, because once you set those items up once, they are there for you to use over and over.

Bid Item Numbering

You can set up your bid items however you want as long as it makes sense for you and the client, but I recommend a logical numbering system. I start off at 1000 for my first header, which for the bids I create is always erosion control. I start them off in the one thousands. I leave a few numbers between each bid item for flexibility. So for a bid I might set up, my first bid item will be 1000 for silt fence, then the next one will be 1005 for erosion hills. I like to leave five places between each one. That way if I need to add something it's not going to break the system. So leave room for items that may need to be inserted between your standard set of bid items.

You may use the same bid items for five years. But then you pick up one client that says, hey, on this one particular job I need you to add this, this and this. So then there are three or four more bid items that you've got to include that you normally don't. After this particular job, that may be something you make your new norm. You may want to bid on that stuff on a regular basis, especially if you're able to show a profit from doing it.

If the Client Requests Changes to the Bid

If a prospective client requests changes to a bid, you just have to go in and edit your bid to account for the requested changes. If you're not showing extensions, or your actual unit prices, then a lot of times you can give back the cost of that specific item but still keep the profit you had figured in the bid for that item. That way you've got a little more incentive in your bid and you are seeing some return because the client has asked for something in return. It's kind of a win/win for those situations. Just don't get greedy, to the point where you can cost yourself a job if you keep too much.

You should also strive to present your bid in a very professional looking manner. Have some kind of logo, have everything neat and orderly, and make it easy for the client to understand what you are proposing.

I have a good friend named Robert Spickard who is probably one of the greatest men I've ever met. He's in the drilling and blasting segment. He once told me: everything you have - your equipment, your pickups - they don't have to be top of the line - they don't have to be brand new - but they should be neat and clean. It's the same way with your bid.

You want your bid to be neat, clean and understandable. There is no worse feeling than giving your bid to a client and then having them call back and say, well, I don't understand this, I don't understand that. You want it to be cut and dry. You want them to be able to look at it and not have any second guesses on what you are actually bidding on. Not to beat a dead horse, but having a system in place that lets you create a professional quality bid from a standard set of items will save you a ton of time and very likely help you win some jobs.

Chapter 4 – What to Charge for People and Equipment

So you have just started a brand new construction company.

Congratulations.

Now you're bidding on a job. You have no idea what you should charge for your people and equipment. How do you figure that stuff out?

Let's start off with people. There are a lot of things you've got to take into consideration. You've hired a man and you've agreed on a salary or hourly wage you're going to pay him. A lot of people will let that be it. They think: that's what I'm going to charge for this guy, plus markup.

Sorry, but if you hire him at $15 an hour, he's actually going to cost you about $25 an hour and maybe more, depending on how many perks you offer him through your company. One thing is Workman's Comp. Here in Tennessee where we're at, it varies but, for the most part, you're going to be looking at somewhere between $7 to $9 an hour that you're going to add on for Workman's Comp insurance.

Now you're up to $24 an hour for that $15 employee. If you furnish him a vehicle and gas to drive back and forth to work, you've got to include that. What's that worth? You've got to figure out what that fuel or what that truck is worth per hour to add onto him. Then if you furnish him a phone, if you furnish him insurance, all this has to be incorporated. This guy that you're paying $15 per hour could cost you $30-35 an hour.

Every employee is going to cost you more than what his hourly wage is. There's no way around it. You've got to do your research. You've got to talk to your insurance company to know what insurance is costing, to know what Workman's Comp is costing, and figure out how to break it down per employee.

Another thing you've got to figure out is what your overhead is going to cost you, and that goes back to your equipment, your employees, and everything else. All this goes in on a bidding level. You've got to figure out what your actual cost is and be able to break it down to know if you're making money or not. Going through this process will help you figure out a rate you should charge for your overhead. It may be 5%. It may be 10%. I can't tell you a magic number, but you've got to come up with that number. Whatever the rate is, you need to figure it out and make sure you add it to your bids.

In addition to your overhead, you've got to decide what you want to make as far as profit. You don't want to be bound by a particular profit rate to where you never get any work. You don't want to be greedy, but you want to be profitable. If you've never added profit onto a project, a good starting point is 10%. Roll the dice. Get your first job out and, if you can show profit after that 10%, you may want to carry on with that. If you show a loss, try to figure out where that loss is at.

This is where it is tremendously useful to use a cost tracking tool that will allow you to match your costs against specific portions of the job to know exactly where you lost money and need to make corrections. A 10% profit level may be all you ever need, but 15% may be your magic number. You don't know. You've got to analyze your numbers and figure that out.

Equipment is a little different. There are a couple different ways to look at it but I will tell you how I like to do it. Most of the equipment that you purchase is going to depreciate in four to five years, or it's going to get to the point where it's going to need some major maintenance or even be replaced. What I've always done is take the brand new cost and add 15 – 25% to that amount because that is what it will cost to replace in five years. So knowing I've got to account for future replacement at a higher cost than today, I figure that higher cost into my rates.

Another thing you've got to account for is maintenance. If you need an undercarriage or your engine fails, you need a new transmission – it could be a lot of things that could cost you a lot of money. There are 2080 working hours in a year, so that's a good place to start. You take that equipment cost and you divide it by five years and then you divide it again by 2080 working hours. That tells you what you've purchased that machine for, per hour.

Another thing that you've got to figure in is the cost of fuel. If that D6 dozer burns 80 gallons of fuel in an eight-hour day, and your fuel is costing you $2 a gallon, it's pretty simple math to see what your spending in a day on fuel. Divide that total by eight and that will tell you your hourly fuel cost – your fuel's costing you $30 an hour, or whatever it may be. You've got to add that to your equipment cost.

Once you've added the cost of your operator in there then you've got your margins set for your markup. Using a bidding and costing application like ProfitDig will automatically mark up the items in your bid for your overhead and profit. If you're not using an application that applies your markups for you, then you will need to do it manually.

So with equipment, just like with your employees, your actual expense is way more than just what you are paying the bank or equipment company. There's a lot of stuff that has to be stacked on top of that to really have a true cost of what your equipment is costing.

Here's an example of the kind of thinking you should apply to the decision of how to cost your equipment. Some companies use this strategy all the time.

Say I just closed a job for which I know I'm going to rent equipment. Well, we'll go back to our CAT D6 dozer for this example. I go to my local CAT dealer and I say, "You know I'm going to need a D6 for a month. What's it going to cost me?" He may tell me $8,000 a month, plus tax. Well I know I'm going to be working roughly one hundred and forty hours a month so I just take that amount plus tax and divide it by my hundred and forty hours, which gives me a per hour rate I'm paying for that machine. At that point I've just got to figure in my fee and my operator time, and put a mark up on it. If I end up using my own equipment, I'll be making really good money for use of the equipment because my actual expense is going to be at a lesser rate. If I still have to rent the equipment, I'm covered because I have based my rate on the rental rate.

If all of your equipment is being used, and you know you've got to rent, I suggest a pricing strategy like the one above. If you've got equipment sitting in a yard, you're going to go with the lesser of two evils. If you are using your own equipment, you can use your actual hourly rates for your equipment and have a better chance of getting a job. That's the only downfall I see to using rental rates as your basis for your hourly equipment rates. It may limit the jobs you can pick up because your bids will be higher. Where you might get five or six jobs out of ten you bid using the lower rates, you now might only get one or two out of ten because you're bidding at a higher level. You're also going to make more money on each job. It just depends on what your strategy is, and how much work you've got on the books.

If you have equipment sitting in your yard that you can charge a rental rate for on a job, you're going to make a lot more money on that equipment.

Chapter 5 – Contract Negotiations

Let's talk a little bit about contract negotiations.

It's pretty common in the bidding and contract awarding process for the general contractor on a job to call and say they need to negotiate your bid. Most of the time, if you're bidding as a sub, it's always private. The G.C. may be bidding on a public job, but the deal that you have as a subcontractor with that G.C. is going to be private. You will likely have no idea how many other contractors - other subs – you're bidding against (it could only be you – but you won't know that!).

If you get a call from them saying they need some help, or saying that you're too high but they want to work with you, you've got to be prepared to negotiate.

And it's not always the case that you need to, or have to, reduce the amount of your bid.

The biggest part is going to depend on your workload. Are you at full capacity? Are you struggling to find men to do the work? Are you trying to juggle crews here and there to keep up with everything you have going on? If you are in that situation, then you've got to decide if you want to take on the job to keep growing your company, or if you should pass. If you are already stretched, I would suggest opting out of negotiations. You need to have good money in the job.

I would probably tell the G.C., "I would like to do this with you. I want you to continue to send me bid invites, and give me the opportunity to bid work to you, but this is the lowest number I can give you - what's on my bid."

It might cost you the job, but not necessarily. It's kind-of a gamble at this point. Does he really have other quotes from other subs? For all you know, you may be the only quote for the scope of work that he has in front of him. It may be a game play, with the G.C. trying to increase his profit margin.

If the G.C. tells you that he has other lower bids, that may not necessarily be the case.

Most of the time, they are going to have at least two or three other subs bidding on the same scope of work that you're bidding on. But the G.C.'s are busy - just like you are - and they may have only taken a bid from a sub they've used in the past and know they can trust. When you submit that bid, they may call you back and try to increase their profit on your part of the job. So they will say they want to negotiate. They may tell you they have lower numbers than yours, but do they really? You don't know. That's just a gamble that you're going to have to be willing to take.

What if you really need the work?

You need to think about that how badly you need the work early in the bidding process. I'm not saying that you want to cut your numbers early on, but you need to have some kind of contingency in your bid, in case you do have to negotiate.

What I like to do is add 5% contingency to my bids. If I do not have to negotiate, then I've got that 5% out there for "unseens" – unexpected things that may come up, that I can't get a change order on, or that I've missed in the bidding process. If you've got that extra 5%, then you're covered. If you do have to negotiate, you've got your costs, your markup, your profit, and your overhead all covered so you can negotiate up to 5%.

As a general rule, as a subcontractor bidding on work for a general contractor, you need to allow yourself some leeway, like the 5% discussed above, to account for unexpected expenses or to give you a little bit of room to work on the price of the bid.

You don't want to go into the bid with a hard number that cannot be altered in any way, because even if you desperately need work, you don't want to give up your profit. And you don't want to give up your overhead to get that profit. You want to give them something back to show that you're willing to negotiate under certain circumstances, but you still need to make money on the project.

Chapter 6 – Contract Review

So you have just bid on a job, or part of a job as a subcontractor, and you've received a contract back from the general contractor, or whomever you're going to be working for. It is very important to review the contract that has been sent back to you, especially if it's a company that you've never worked for. No matter how busy you are or overwhelmed at the time, it's to your advantage to take the time to read through that contract, and make sure that it's everything that you bid on.

I'll give you an example of the kind of thing that might be different from what you get back as opposed to what you bid.

Let's say you have submitted a bid to put in a waterline that sits in the middle of the street. Most all contractors and subcontractors are going to qualify their bids with a list of exclusions and qualifications. On this particular job, you have included the labor, materials, equipment, traffic control, signage, whatever you need to put this project in, but you have excluded pavement. You are not a paving contractor. You do not want to deal with the paving issues or have to deal with a sub-paver under you, so you exclude pavement in your bid.

Then you receive a contract back, and as you're reading through it you notice that in this contract they have you listed as including all pavement patch for your scope of work. Well, you did not include that in your bid, so you don't have money to cover it.

In the process of reading through the contract, you're going to strike through that, or highlight it and initial it, and any other discrepancies you may see, and then send it right back to the G.C., or whoever you're contracted with on this project, and say, "This is not part of my scope of work." At that point, they will correct it and resend it back to you...or you're at a standoff.

At that point, you need to be able to talk to the contractor that you're potentially under contract with and work it out. If you can't, you may want to take the contract to your lawyer, let him review it, and get suggestions from him.

Do not think that it is unusual or unprofessional to strike through items in the contract, initial it, and send it back. In almost every contract I have worked on, I have had some issue that does not pertain to the bid I submitted, or something that stands out to me as fishy. 98% of the time, the other party will realize that there was a mistake and agree to it.

Most of their contracts are boilerplate. They're just standard issue contracts that they issue over and over, with every sub, or every other contractor that they are dealing with. So if they see that there is an issue there, they will go ahead and correct it and send it back to you. Then if you both agree on it, you both sign contracts and everything is good to go.

If you go through that process, and the general contractor, or the employer on this project, does not agree to your changes, then you're going to have to come an agreement one of three ways - he's either going to have to issue me a change order to incorporate this into my bid, or he's going to have to make his changes to take this out of my bid, or we're going to just annul the contract and pretend like it never happened. Then the other party will go on to his next bidder.

Until you sign it, you always have the option to decline the contract. You can kick it back and say, "I'm not interested. You'll want to find someone else." You can go to your lawyer, and see how are they going to help you - "Can you write this contractor a letter, explaining to him why we're not going to do this portion of work?" Show him a copy of your bid.

You can also ask them to include a copy of your bid, with your exclusions, and your qualifications, as an exhibit – exhibit A, exhibit B, whatever it may be. They may already have an exhibit A, B, C in your contract, but you can ask them to incorporate your actual bid as an exhibit.

The main point is to carefully review the contract you get back from the lead on the project, and make sure that it matches up to what you actually bid to do. It's pretty common for smaller contractors to kind-of rush through things and not really take the time. There are some situations where it is more difficult, like if it's a job that is larger than what you normally do, and has more items than what you typically bid on. Whatever the circumstances, you really need to be aware, and take the time to read through and understand the contract.

A lot of general contractors will have an excessive contract and sort-of hide stuff here and there. It may be an 80, 100, even 200-page contract. I think they do this on purpose. If you miss something, later on they can hold your feet to the fire, and say, "You are going to do this. It's in the contract."

Read every page of the contract and make sure there are no discrepancies between what you bid to do and what the contract says you will do.

Chapter 7 – So You Won Your First Bid – Now What?

So you've got your first successful bid under your belt. Now it's time to start your job and start performing. Starting with your very first job, you've got to control your costs. Not just your first job, but every job. This is going to give you your baseline or rule of thumb for your bidding from this point on.

You're going to have to track your costs every day. When you bid a job, you figure X amount of dollars for labor, X amount of dollars for your equipment, X amount of dollars for fuel, and X amount of dollars for materials. You need to track your costs to know if you are on point with your bid.

There may be some portion of your bid that you can't perform yourself, and so you've had to figure in for a sub. For your sub-contractors, you will have a quote that gets figured into the bid. You're going to mark them up just like you do everything else because you want to make money off them as well.

Take using a sub for road boring, for instance. Your sub that you're using to do your boring may perform the bore itself, but you've got to supply the materials for it. You may have to supply the machinery to excavate their pits, set their machine in and out every day, and set their pipe and other materials in and out for them. You may have time and equipment associated with that bid out as well.

There are a lot of different scenarios, but having your budget to work off of that you created with your bid, and then tracking your actual cost against that on a daily basis, is going to determine if you're where you need to be in your bidding process or if you need to tighten up and get a little better at it.

Every day you need to know how much time your men spent on the job, how much you spent on fuel, how much you spent on materials – everything. It must all be tracked every day.

Let's say you and I are bidding on the same project. We're two different contractors. We both should have a good working relationship with our suppliers. Our materials for that project should run within 5-6% of each other's. My materials are going to cost me the same thing your materials are going to cost you. How quick I think I can get the job done, and how quick you think you can get the job done - that may be the determining factor if I get the job or you get it. It's going to cost me money, just like it does you, to have equipment out there, buy fuel, pay my men - that's where I've got to be a better estimator.

If I think I can do this job in X amount of days, and that's how I base my bid, then I've got to make sure this job happens in X amount of days or less. Underestimating a project is where you can really lose yourself. You want to be as quick as possible to try to make as much money as possible, but you also want to be smart enough in the bidding process to give yourself enough time to actually do the project.

How do you do track your costs?

It's tough - especially if you're the owner and you're actually out in the field doing the work. It gets really tricky trying to keep up with all these receipts, your materials coming in, payroll going out, while working 10-12 hour days in the field. There's not much time left and you don't want to take away from your family time.

You don't want to get to the point where you and your wife are on the outs because you're always working, not spending enough time with her or the kids. Those are things that most people in our line of work won't even talk to you about. You've got to figure that out on your own. I've been there. I've come to blows with my wife - about lost my wife a few times. You need to make sure you're taking care of your home work at home before there is no home work at home.

And that's why you need some kind of system that simplifies the process of keeping track of everything.

There are programs that you can buy for a lot of money to do this for you. But you can also use a low cost, monthly system that will be like a drop in the bucket. I recommend ProfitDig, a product that my partners and I at ProfitDig LLC spent several years building.

With ProfitDig, when you create your estimate, you also create a working budget. It's mobile friendly. If all you have is a cell phone, or smartphone, or laptop, or tablet - whatever you have in your truck - in just a few minutes time before you leave the job you can enter how much time your men have, how much time your equipment has, how much fuel you used, etc. Say you had a fuel delivery today. You can enter that fuel ticket in at the end of the day, along with other expenses, and your work's done for the day.

Let me repeat, because this is very important: Before you leave the job site, you should log your expenses into some kind of system. Record that these guys worked this number of hours, we spent this much on costs, this much on fuel, this much on whatever. It goes into the system, and at the end of the day, as long as you do that before you leave the job site, all of that information is recorded. You will know in real time how the project is progressing compared to your estimate.

So it's the end of the day. You're the crew leader. You've got X number of men working under you. You know every day it's going to take you an hour, maybe 30 minutes, to get your job site cleaned up, safe and ready to leave. You can put your crew to doing that work. Meanwhile, you sit in your truck and spend 5 or 10 minutes time entering your man time, materials, fuel, whatever you need to enter, and get that done before you ever leave the site. That will be the best 10 minutes you could ever spend because it will keep you on track.

When the job is done, if you have recorded all this information every day like you're supposed to, you will actually know - did you make any money? Did you lose money? How much did you make or how much did you lose?

That will certainly affect how you estimate future jobs, because you'll be able to bid more accurately.

Chapter 8 – Job Management

If you're the owner of the company and a bid has been accepted for a job, you can just sit back and take it easy while your men do the work, right?

No, because that's where most of the work really starts. It takes a lot of work to prepare the bid and to get the job awarded to you, but once it is awarded, as the owner or project manager, then it's your job to oversee it and make sure that this job comes in on time and on budget.

There are a lot of steps to this process, but step one is to go over the contract, like we discussed earlier. At that point, you as a project manager or owner have to know exactly what is in the contract. Somebody has got to read through the contract and make sure that you're protected because you know the contract terms are going to benefit the company you're working for. They're protected. It's their contract.

So the first thing you need to do it read through, look for any loopholes that you may see, and determine if this is a contract you feel comfortable signing. If it's not, then there will have to be some kind of negotiation between you and the company or entity that you are going to be contracting with.

Eventually, assuming all the contractual stuff gets worked out and you sign the contract, you start the job.

Now you have to start managing. You've got to manage your equipment, your labor, and your materials. You have men out there who are performing this work. As a project manager, you've got to watch over them, and watch the production they're getting because you have bid the project with an estimated or proposed amount of production per day or per week or per month. So your production has got to be pretty close to what you estimated for the job to make money.

You have a time limit on every project. You've got X amount of days. If it's a big project, you may have a year or a year and a half to complete, making it even more important to keep up with the project on a daily basis.

The situation is even worse in a bigger project because as the project manager, you've got a lot of stuff going on. You may think, "Well, we're not getting quite the production, but I don't have to worry about it right now. As we get a little more into it, we can probably pick it up somewhere." But sometimes that's not necessarily the case. The job may run slow all the way through, and if you don't do something on the front end to correct it, then you may end up paying liquidated damages because you're so many days behind.

As far as your exposure to damages, it varies from job to job. It could depend on who you're working for and what kind of contract you signed. You may not have read it far enough along to see that you have $1,000 or $1,500 per day liquidated damages. So if you're three months behind, that could put you out of business pretty quick. That is why you must read through all of your contracts carefully.

As a job goes along, you have to pay attention. There is always going to be stuff to manage as far as documentation. You've got to watch over your field notes as they come in, so that's one thing that you need to make sure that your crews or your foremen and your superintendents are doing very well. I hate to say it, but sometimes you end up in a lawsuit, and the better documentation you have, then the better off you are in court.

You need to make sure that your men are giving you adequate notes about what's going on every day. If someone on your staff has a conversation with the project manager or owner you're working for, that could be something to come up. He needs to document that that, that and that came up in the job today. Someone said this or commented on something we're doing. That type of thing could come back to bite you in the long run if it's not documented.

Often the requirements of a job will change. It may be something that you didn't see or the engineer or the owner didn't see when this job was originally designed or planned. At that point, you're already contracted with them, so they're going to come to you and ask you for a change order for this particular item. There again, as project manager, you've got to be on your toes at all times because there is the potential for it to be a very profitable change order.

Typically, your GC, the municipality, or whomever you're contracting with is not going to bring another contractor in on top of you. You're already there, you're mobilized, and you're working. At that point, you have the possibility of squeezing out some more profit. You might have originally bid the job at 20% profit, but decide to try for 25% or maybe 30% on the change order.

But once again, it might all depend on the contract you signed. Some GCs that you contract with will have it in their contract that you can only do a 5% profit and a 5% overhead, or maybe it's a 5 and 10. Then you've got to give a breakdown of your material costs, your labor costs, and equipment costs. But there are ways around that.

You've got to be creative as a project manager. What I do in those situations where I have a very restrictive contract is to adjust my estimated hours to give me some room. If it's going to take me 10 hours to do this change order, I may put 15 hours in there. If I know it's going to take me 2 loads of stone, I may figure 3 or 4 loads of stone. There are ways to pick up extra money, but as a project manager, you've got to be able to see when those things are needed and what you can do to give yourself some flexibility.

Here is an example of using a change order to increase profit while also saving the customer money. I did a job several years ago in which we had a massive box culvert that went between two roads and around the back of some properties. This particular part of the bid was over $400,000. I did some value engineering, took it to the engineer for this project, and said, "Hey, I think there's potential savings here for the developer and it could speed this project up as well." He was interested, so I took my notes and the drawings I had drawn up, and I went back to him and showed him what I had.

What we ended up doing is keeping the box culverts and wing walls in place on each road, and then we actually cut in a drainage swell on the property line in between the properties where the box culvert was going to go and placed Rip-Rap on the bottom and banks for stabilization. I can't remember exactly how much labor we had estimated to start with, but it went from several weeks' worth of work down to hardly any time at all. We gave the customer around a $300,000 savings, which made him very happy. As for us, we had bid 18% profit on the job, and when it was all said and done, we ended up with 42% profit.

When I showed the owner the changes and what the savings were, all I did was give him back the cost of the labor, the cost of the equipment, and the cost of the materials. I had the 18% profit in my 400 plus thousand dollars. I kept that separate. I kept that for myself. So I figured up what it would cost me just like if it was a totally different job, to put those two box culverts in on the road and cut that swell in, and then that determined my markup. I added my profit from the box culvert to that, and that's what I gave them back and it ended up being around $300,000 savings.

That is the kind of creative thinking that can dramatically increase your profits.

As a side note, I never did receive a bonus I was promised as part of that deal, which is why if you want to make the big bucks, you've got to work for yourself.

Digging for Profit: Secrets to Success in the Construction Industry

Chapter 9 – Job Costing

"When you put it like that, it makes complete sense."

So you've started your company, you've bid on some jobs, you've done a few jobs - now what?

Hopefully you have engaged in accurate job costing and seeing what money you actually spend and what it costs you on a per bid item basis. You will know these things if you have been using a system to capture that information daily. You will also know if you have a leak in your company. If you're losing money, you know where that leak is.

It's not just about knowing if you're making money on the job, it's about knowing what part of the job you're losing money on.

Every part is important. You figure profit on each separate bid item that you bid, and whether your profit is 10% or 20%, you have figured X amount of dollars of profit in the job for each one.

Let's say you figure 15% profit across the board, so you've got 15% profit in your total job. When that job is over and your job costing is done, and you show 12%, you're might think, "well I made 12% on this job. I didn't lose any money." Sorry, but yes you did because you originally figured on 15% profit, so you lost 3% of what you thought you were going to make. So there's a leak somewhere. You've got to figure out where it's at and get it stopped, and continue to make the profit level you want to make. 3% can make a big different in the future success of your company.

Now as markets vary, your profit level is going to change. When markets go down, your profit level is going to go down. But again, when you've got a few jobs under your belt, you've accurately tracked job costs, and have identified your leaks - some stuff you've overbid and some stuff you've underbid – then you can find a balance. You know where you need to be, especially on similar jobs to what you've already completed.

Through a system like Profit Dig, you can look back and see exactly which bid items cost you money. Say it's moving dirt. Say you've got 10,000 yards of dirt to move. You think you can move it in five days. Well, it actually took you seven days. That may be your hole. That may be where the problem is, but without knowing that, then you have no idea. You will likely keep making that same mistake on your estimates, and continuing to lose money in that area.

Every job is going to vary, but at some point, there are going to be a lot of similarities, and so the next time you look at a project that's similar to this one, you might say, "Hey, you know, I thought I could move 10,000 yards of dirt in five days. Well, this time I'm going to figure eight or nine days to make sure I'm covered."

Better job cost tracking leads to better estimating, which leads to more profits.

Chapter 10 – Managing Materials

Let's talk a little bit about materials.

You need to have relationships with multiple suppliers. After you have worked a few jobs, you can look back and compare the quotes you have received from different suppliers and see what your potential savings are working with one supplier or another for different materials. When you contact a supplier for a quote, try to get a feel for how bad they want the project and see if they are willing to give any.

You are not trying to cheat anybody or do anything wrong. You are just trying to get the most profit possible out of your project, which is the job of any business owner. If a supplier is going to give some, then he will. If he can't, then he won't.

In my experience, I have been able to pick up 5% to 15% savings just from making phone calls and saying, "Hey, can you help me out on this project?" If it is a substantial buy, then a small percentage break in price can make a big difference to the bottom line. Say you need $200,000 in material - a 5% to 10% savings can mean $10 - $20 thousand in additional profit - a pretty good savings upfront before you even start the project.

A lot of times you want to try to build relationships with a certain supplier. That is not uncommon. A lot of companies choose to use one particular supplier most of the time. In having that type of relationship, you would hope for the cheapest price upfront. A lot of times, you can call that supplier and say, "I need a little bit of help." If you have that relationship built, then they are usually more than willing to give you some money back.

But ultimately you have to do what you have to do to get the best price on materials.

Once you have ordered the materials, you need to also think strategically with regard to things like how much of the materials to get at once and where to put the materials on the jobsite.

A lot of construction managers overlook this, but your suppliers are going to charge you delivery charges. You want to try to get full loads every time to take advantage of that, because you want to get as much material as you can for that one delivery charge. If you need just a handful of material, the supplier is going to charge you the same delivery charge for that as for a full load.

It is important to make sure that you have room on your project site to store your material. A lot of times projects are tight as far as workspace and you can't handle more than one load at a time. If you're lucky, you can handle two or three loads. It may be important to strategically order your material in phases, as you need it to optimize your delivery charges.

Proper material placement is up to the foreman on the project. He has to be conservative about where he is going to put his materials and where he can hold the most materials. It needs to be strategically placed so that it is easy to get to, with easy access in and out.

One of the worst things you can do is stockpile your materials and then find out a few days later that it is in the way. You end up having to move it, wasting labor, wasting equipment, and wasting fuel. These people working for you could be doing other things to move the project forward, but now they have to move materials because you didn't think it through and put it in the wrong spot.

In my opinion, I would say that stone is one of the most wasted materials on a job site. The key to making good use of your stone is to be proactive in conserving materials. Before you order the first load of stone, you need to figure out exactly where you want to put it. Tell your loader, "Hey, we need to get an area cleaned off. We need to have a good smooth surface to stockpile this stone."

I have seen it happen too many times. Stone shows up and nobody is ready for it. You put it in a wet, muddy spot because it is open. But you also lose 10% or even 15% of your stone in the process - every time you do it. Maybe the spot is full of ruts. You can never get that stone back out of those ruts.

You also want to think about ease of access. When you pick the spot where you want to stockpile stone, you don't want to put it in the furthest corner away if you can help it. You don't want your loader man to be wasting time driving across site to pick up stone when you could have had a central location for easy access from all points of the job.

So put stone somewhere where you have easy access but also where you minimize how much of it you waste.

Say you are doing utility work. You know that it is going to take you one load of stone per 100 feet of pipe. I would try to pick a dry day before we start the project, order my stone that day, and then space it out every 100 feet. If that doesn't work or isn't possible, try to do two loads every 200 feet. Try to get to where your loader can choose from one side or the other, wherever there is quickest access to keep production up and speed your project along.

Managing your materials can have a big impact on the success of a job.

Chapter 11 – Equipment & Labor

"There's something about seeing red that just drives me crazy."

Now let's talk about your equipment and your labor.

If you have bid a job accurately, then your bid tells you how much labor and equipment you need. Every project manager or estimator who bids a project has in mind what it's going to take - how much labor, how much equipment, how much fuel - to put a job in. Now, does it always work out like he planned? No. But he's ultimately responsible for the profit and the budget for that particular project, so that bid tells you where to start.

It is very easy to have too much labor, and the best way to notice this is a two or three day observation. In two or three days' time if you've got labor standing around with nothing to do and you can't find anything for them to do other than busy work, then your labor heavy. You need to relieve some of the labor.

It is the same thing with equipment. If you've been working two or three days and you've got two or three pieces of equipment that have not moved, and more than likely will not move for a while, then it probably doesn't need to be on the project. That equipment is there to be used to make money. If it's sitting still, it's not making anybody money.

So a lot of it is just observation. Is your labor and your equipment being used?

If you see that you have too much equipment or too much labor for a job, then it's going to cost you money. For one thing, you wasted money by having that equipment hauled in there. Now you need to get that equipment moved to a job so it can be used to make money. That means you've got to waste another mobilization to move the equipment back out.

If you have unused labor, you've got to get them on another project to relieve the stress from this project and try to utilize them in a way that could be beneficial and return a profit on what you're paying them.

If you don't have any other job to send that labor to, at that point it becomes a personal preference for you and your company with regard to your relationship with your employees and your equipment. If you don't have anywhere for it to go, your equipment could be in the way. At that point, it's not making money one way or the other, so you should have it moved back to your shop, yard, or wherever you keep your equipment sitting when it's not being used.

As far as laborers, if you've got the resources and you want to try and keep them on and not lose them, you can keep them on the project and take a loss. That's a personal preference, but if you want to make the job profitable there may be a point in time to where you need to do a temporary or permanent layoff until you can do better.

If it turns out that you don't have enough manpower or equipment to do the job, then you will need to pull in resources from other available labor and equipment from the company, if it is there. If it's not, then at that point you need to look for labor outside the company.

Over time you will notice certain leaders in your company - your crew leaders, your foremen, your superintendents – who will call you and say something like, "Hey, I've got too much labor. I've got too much equipment," or vice versa, "I don't have enough." Those are the managers you're going to recognize. Those are the ones who you're going to want to make long-term people for your company, because they're watching out for the best interest of the company.

It all comes down to personnel. It's who's making the best decisions for my company. As a project manager or superintendent, it's my job to oversee these projects and these people and make sure they're doing the best they can possibly do; to make sure our resources are being used wisely and not wasted. But you don't want to overextend them either.

Chapter 12 – Rental Equipment

Rental equipment can play a big role in the success of a job. You get to the point to where, as an estimator or project manager, you should know what your company's resources are, what's available, what equipment is coming in, what is currently being used and not being used. If you're running at seventy to eighty percent capacity and you're still bidding work, there's a pretty good chance that you need to start figuring in rental equipment rates for your bids you're turning in.

If you have a rental equipment company you work with primarily, or multiple rental equipment companies you work with, every so often you should call them and ask for updated rental rates on certain pieces of equipment, or even all of their equipment. I like to do this every 90 days. It doesn't matter if you use the smallest piece of equipment or the largest piece of equipment, you should have those rates and know at all times what it is going to cost you to rent what you need.

You need to break the rental rate down into hours for bidding and job costing purposes. Most of the time, you're going to be working a twenty-two day month, or roughly five days a week, and if you're working eight hours per day, or nine hours per day, it's just simple math. You figure out what that rental rate is plus tax plus insurance.

And another thing: Either you can add that equipment onto your insurance rate, which is a cost, or you can use the rental company's insurance rate, which is usually higher than what you're paying. You figure those costs and you divide it by your number of days and by your working hours, and that gives you an hour rate to use.

You also have to put some kind of mark-up on there. If you generally like to use 10% profit, or 15% profit - whatever your profit margin is - you need to add that onto the rental equipment rates because you want to make money off of it as well.

Another important thing to remember is that your insurance company does not know when you turn that piece of equipment in, so it is very important that your project manager knows when you send a piece of rental equipment back. He (or you) will need to call your insurance company and say we no longer have these pieces rented, so I need to take this off of my premium. If you don't, then you're going to continue to pay for that insurance. It may be two, three, four months down the road before you realize it. That is money wasted.

What type of equipment you will need to rent depends on the nature of your business. If you're in the landscape business, you're going to need skid loaders and mini-exes. That's equipment you're going to be looking at and can rent from any rental company. If you're doing mass excavation, you're going to want big equipment. You're going to have to pretty much skip the rental stores and go to your Caterpillars and your Komatsus – your bigger equipment companies – the actual dealers that also have rental equipment.

If you see your company growing and can use this equipment down the road, or you've got a year-long or longer project, then a lease purchase might be a good arrangement for you. On a lease purchase, most equipment companies will give you 90% to 100% of what you paid on that equipment in rental payments as a down payment. At that point, it becomes your own piece of equipment. Your payments go down, and you can utilize it at a cheaper rate.

Getting the best rental rates is another relationship situation that you build over a period of time, but again, it's always good to get multiple quotes. You never want to get just one quote. You want to get two or three quotes so you can make the most money by getting the best rates. That's the name of the game – to be able to utilize your resources and to get the most out of your project that you possibly can.

If you know that your company is running 80% or better as far as equipment and manpower that you have, and if you think you're going to stay there for a while or you possibly could be growing larger, at that point I would recommend that you go ahead and make the purchase, even if you have to finance it. Your finance charges are going to be fifteen, maybe even twenty percent cheaper than what you're paying in rental rates.

The caveat, obviously, is that once you sign that note, you can't really turn that equipment back in. With rental equipment, if you get into a tight situation, you can just send it back and stop those payments. So it's a matter of weighing getting the best rate (buying) versus having more flexibility (renting). You have to make that call based on where you are as a company and where you expect to be in the future.

Chapter 13 – Subcontractors

There are going to be times when you can't self-perform every part of a project yourself. It's very common to be looking at a project, especially something you really want to bid on, that is above and beyond what your capabilities are. That's where subcontractors come into play.

Say, for example, you're an excavation and utility contractor bidding on a job that has paving in it. You're not in the paving business, so at that point you've got to seek out a subcontractor for that part of the job. You've got to find a subcontractor that you can build a relationship with, one that you can depend on, and one that's going to be fair as far as pricing and working with you.

Specialization as a subcontractor, performing one part of a job, is a good way to get started in the construction industry. I would say most general contractors probably started out as subcontractors doing specialized work. They start off either in the building or structural industry, home building, excavation and grading, or maybe utility work or paving.

61

Some companies choose to stay specialized, which can be a good way to go. There are a lot of contractors out there that are strictly paving only. It's whatever you have a passion for, whatever you think can be the most productive and most profitable. At a point in time, that may lead into something else because they all kind-of go hand in hand.

The job ties everything together. I may one day be looking to be that GC that can do it all. I can do the excavation. I can do the paving. I can do the structural. I'm a one-stop shop. But you've got to start somewhere, and that's the way most companies start out - in some kind of specialty line.

Say let's say that's the approach you've taken, and you do have a specialty. How do you go about making it known to bigger companies or general contractors that you are a subcontractor and this is what you do?

You need to make it known to bigger companies or general contractors what your specialty is and that you are available for work. If you're looking at companies to work with that are primarily in your area, the best thing to do is get some business cards, and on a good rain day when you can't work, ride around town, pop in their offices and say, "Hey, I'm so and so and I am a subcontractor. This is the line of work I'm in. If you ever need my services, here's my contact information. I would really appreciate you giving me a shot."

On the other side, if you are the general contractor, you typically will have built relationships over the years with particular subcontractors. These are the subs that you rely and depend on. If you're needing somebody else that you haven't dealt with, you can look in the phone book, or yellow pages, or consult with the Better Business Bureau. There is social media now. Many will have their own websites. There are a lot of different ways. If you look hard enough, you can find a sub for whatever particular specialty you're looking for.

One thing to keep in mind with regard to subcontractors is that they are subject to the same safety qualifications and regulations as everyone else on the job. Most of the time these days if you're working on a job site you have to have a certain amount of OSHA training - maybe a ten hour, twenty hour, or even a forty hour course. If those qualifications have to be in place for this particular job, as a GC looking at a subcontractor, you've got to make sure that he has that proper training. If not, then you need to make sure that he gets that proper training before he starts his project.

The last thing you want to do is get in with a developer or a project owner and hold the job up right off the bat because you don't have the training, or your sub doesn't have the training in place that he has to have. You could end up losing a potentially good relationship.

Once you hire a subcontractor, then, he's part of your unit at that point – at least for that particular job. You hope to build a relationship with him to where you use him on multiple projects.

As a subcontractor, you want to build those relationships to where your employers know you're dependable, know you're going to do a good job, and know you're going to go above and beyond what you have to do to get their business.

Using subcontractors is a big part of the construction business, especially as your company grows and gets bigger. You will eventually have to sub out some stuff that you're not big enough to self perform. It may be a situation where there's a really nice job out there and you want to be a part of it but you're just not big enough to handle everything yourself, so at that point you've got to think about subbing out to another contractor.

Hiring a sub involves a bidding process just like any other job. Your sub, whatever field it's in, has to give you a bid on that particular portion of the project that you may want to sub out. Then at that point you have to evaluate the quote that you get because you want to be able to make money off your subs. So you know you've got to come up with an adequate markup to make money on that portion of the job.

If you're making ten percent profit on a project, you're going to want to make at least ten percent on your sub as well. It's a process you've got to go through to figure who's going to be the best sub to get the job done but also leave you with some profit.

It is also a good thing to have some kind of existing relationship with subs you work with. If you're just starting out and you're just now getting to this point where you may need some subs, it may be a process to figure out who are good subs you can trust and who you know that you can work with. Just like you as a contractor, whenever you bid to another company, it's growing a relationship. You've got to be able to build that relationship with them so they know they can trust and depend on you. It's the same way with you and your subs.

Having said that, you may know someone off the top of your head you know you can trust. That's great, but I still would recommend getting multiple bids to keep everybody in check and to make sure that someone's not trying to take advantage of you.

Through this process, you may actually not make as much money as you had hoped to or even lose some money because you picked a bad sub. If the sub decides they can't do the job for what money they gave you in your bid, then at that point you've got to try to hold them accountable. But you also have to either self perform that portion of the work or quickly find another sub. This situation also puts you at risk of possibly paying liquidated damages and other penalties for not being able to complete your job on time.

So if you're accepting sub bids for a part of a job and one of those subs is one that you have a good relationship with but it's not the most favorable bid, I would you suggest you generally go with the one with whom you have the relationship even if you have a lower cost option. Now if you've got two other bids there besides his and they're considerably lower, you might go to him and say "Hey, you know, I need a little help on this. You don't have the best price, but I want to use you."

If it's a job you really want to get, it's crucial that your subs work with you. You want to come in as close to budget or under budget as possible to have a good shot at getting this project. Subs need to make money. That's the whole point. Everybody wants to make money, but it can't cost you a job because they were too high. But again, in general I would definitely lean towards a sub that I already have a relationship with.

If you are starting off fresh and you're trying to build relationships with subcontractors then I would recommend making a few phone calls. Find out who they've worked for in the past and then give those companies a call. Two or three of them would be good. Say, "How was your experience with this construction company? Did they do a good job?" Word of mouth is always going to be your friend when it comes to finding companies you can trust.

If you are going in blind, and you've got a sub that you have never dealt with and can't find any information about, then you've just got to do the best you can at making a judgment. You can tell pretty quick when they come in if they are conscientious about their job and how they represent themselves. As discussed earlier, neatness counts. If a sub you are considering appears neat and safety oriented, that's a pretty good sign. Obviously, any sub you are considering should have a strong work ethic and adequate manpower and equipment to complete the job.

You want someone who's not going to be searching for problems or giving excuses as to why he can't be there. Every job site is not going to perfect. There are going to be issues, but you want a sub that can work through those and continue to work and not hold you up.

You want to maintain good relationships with your subs going forward. That starts with a contract. Anything you do – from the smallest of jobs to the largest of jobs – no matter how well you know that sub, you need a contract. A contract protects you just as much as it protects them. That's a legally binding agreement that you both have signed saying that the sub is going to do the work that has been designated for the money that's been designated.

You can reprimand a sub, just like you can be reprimanded. If they're not doing a good job, you need to have a clause in your contract. If they don't perform their work like they say they're going to perform, you have the right to kick them off the job and charge them damages for what it's going to cost you to hire somebody else. That's also why you hold a portion of their pay every month so if that does happen you do have a little bit of cushion to pay someone else to come in and finish, or to self perform it yourself if need be.

Chapter 14 – Bidding Strategies

We've gone into some of the technical details of actually creating a job bid, but now let's talk about some strategies that a young contractor might use to win some bids.

Research the Job Thoroughly

I would say one of the first things you need to do is research your project thoroughly. If you're looking at something to bid and you've only got a couple of days to prepare it, I would say let that bid go. You want to look at jobs that give you adequate time to put an accurate bid together. It's for your protection as well.

Get familiar with the job site. Go out and look at it – do some research. If an engineer has prepared a bid package then you can kind-of look at your line items and things that you're going to be bidding on and the kind of work you're going to be doing, and go out and look at the actual site to compare it to. Be able to picture you and your crew or crews putting that site in. Think about how long it's going to take you to move this amount of dirt or put in this amount of pipe. Are you going to have to have subs? How much can you self-perform?

More than likely, your engineer or the owner is going to have soil and bore reports showing what kind of material is onsite to work with. Ask yourself questions like: What is my rock elevation? Am I going to have to have a blaster? Will I do my own blasting? Is the soil going to be suitable to use? Am I going to have to truck in soil or other materials from another site, or possibly even a quarry?

So your first step is to be familiar with the plans, be familiar with the site you're bidding on and what work has be done, and try to figure out the best strategy that you can to get it done in a profitable manner.

You gain a competitive advantage by looking at things with your own eyes versus always relying on what the engineer says. Engineers are humans as well. They make mistakes. They get in a hurry.

Give Yourself Plenty of Time

So again, make sure you've got adequate time to prepare a bid. A lot of times when the economy is good and there is a lot of building going on, engineers will be trying to crank out as many bid packages as they possibly can. There are going to be unforeseen items because they're not going to know everything - they're going to miss some stuff.

As time goes on, you progress in your career, stay in business and do a number of jobs, you're going to notice those kinds of things more and more. That's why I said be familiar with your bid package, know what line items or bid items you have to bid on and see what you have to include. There may be a lot of stuff out there that the engineer left out or doesn't have a bid item for and you've got to incorporate that somewhere in your bid.

Can that hurt you? Yes, it can, because you may be bidding against someone who may not be doing the research you're doing and who may leave some items out because it wasn't specifically in the bid package. If they do, they're going to be low in their bid. They're probably going to get the job, but they're not going to be able to make any money, or they're going to have to turn around and be beating everybody up for change orders.

The kind of diligence we are talking about here can also work in your favor, however. There may be a bid item that you see that you could put more money in because you see a potential for that bid item to grow. Or there may be a bid item that you don't think you need. You might want to put it in at cost. That way, if the owner deletes that bid item you don't lose very much money.

Make Sure the Required Documents Get Completed On Time

Another reason to make sure that you give yourself plenty of time before you actually submit the bid is to make sure your documentation is correct. When you put together a bid package, especially if you're not familiar with the engineer or the municipality or the government or owner that you're working with, one of the first things you should do is I look through the bid package and see what documentation they require. They've got to have your contractor's license. They're going to want a bid bond. They may want a Drug Free Workplace Affidavit. A lot of that stuff has to be notarized, especially your bid bond.

If you wait until the day before or the last day when a bid is due and try to put this documentation together, then it's not going to happen. Your bonding company is not going to issue you a bid bond in just a few minutes' time. It's going to take them a couple of days to prepare that bid bond and get it to you, even if they overnight it to you. So to some extent you need to start working on that stuff as soon as you pick up the bid package. You may have a secretary or somebody in your office who can handle a lot of that for you, but it does not need to be overlooked.

Don't Bid Too Low

Although you certainly want to try to get the work, coming in too low can be to your disadvantage. For one thing, even if you get the job you might not make any money. But for another thing it might even get your bid kicked out if you are too much lower than everyone else. I haven't seen much of it here lately, but some municipalities or some governments will throw your bid out if you are more than 10% lower than the next lowest bid. They look at it like you may have missed something.

Use the Local Contractor Advantage

If you are bidding on certain types of local municipality or government work, you may have a local contractor advantage if the job is in your own hometown. This means that your bid can be a certain amount higher (5% is typical) and you would still be selected over a lower cost bid. So if you're a local contractor and you know that you're going to be competing against some out of town contractors, you've got a 5% advantage. You may be 4 or 5% higher than a low bidder, but you're going to get it because you're within that 5% range.

When you notice a bid package come out and it has that advantage to it, you need to pay attention to it because there's potential to pick up work by using the local contractor advantage.

Of course, if you have a 5% advantage in your local hometown, then you may have a 5% disadvantage if you're bidding out of town.

If you're an out of town contractor and you're really interested in a job and they do offer a 5% advantage, you should go ahead and prepare your bid like you normally would. Figure your labor, your materials, your equipment, your overhead - everything just like you normally would. At that point, you would go back and look and see where in your bid you could cut that 5% and still feel comfortable that you are submitting a profitable bid.

Keep Records of Your Bids

Record what happens in the bidding process. I record mine on a regular basis. In most cases, when you are doing a certain type of work, you're generally bidding against the same contractors year round. I like to keep my bid tabs from each job, and look back at the bid items that I was beat on. Where did so-and-so beat me and how did he beat me?

Then the next time you have a job similar to that one and you know you're going to be bidding against these contractors again, try to figure out ways to get more creative. Maybe you need to put some different people in place to speed up production. Maybe you just need to add more people. Maybe you didn't have enough labor or enough personnel on the project to do it in the time the project required.

If you know a company that has beaten you on a project, when you have a little free time, drive by and check on them periodically. Just kind-of watch over the job, see what they're doing, see how they're performing their work, how they are getting that advantage over you - especially if they beat you on regular basis.

Here is one more tip that works for me. If it's not a real long bid package, I will print off my bid items and I may bid that job three different ways. Then I will make three different copies to read through. Then as I learn more about the project or see stuff as time goes on before my bid date, I've got to determine which one of those bids to submit. I may still have to tweak the bids here and there, but I generally like to have three different ways to bid a project.

That's another good reason to give yourself plenty of time – for massaging the final bid.

Chapter 15 – Size of Projects You Should Pursue

There are all different sizes of projects you can pursue, ranging from projects you can work on yourself right away - smaller jobs in town, for example - to big jobs on bigger sites. As a smaller company just starting out, you always want to strive to grow and get larger, for the most part. But as a smaller company just starting out, don't get too carried away aggressively taking on jobs. I wouldn't try to grow my company by 25% or 30% or 50% right off the bat. The more or bigger jobs you take on, the more resources you're going to have to have.

If you look at something that could be a good potential job for your company, and you may have to grow by 5% or 6%, 8%, it may be worth it.

You should do some research, look into it, and see what it is going to cost you, and what you are going to have to acquire to do this type of project. Typically, it's going to cost more to work in-town than out-of-town.

In-town, you've got traffic issues. It's going to cost you more to mobilize and demobilize, and get your men to and from work each and every day. You're going to be dealing with some kind of Department of Transportation issues. They are going to have control of a lot of your streets, or your local municipality is going to have to cover your streets.

You're going to have different specs you've got to do when you're working in these areas. You're going to have more permits and more fees, which are usually going to be considerably higher in-town than out-of-town. You're going to have a lot more meetings you've got to go to. You're likely going to have to have people to do your legwork for you, because it's going to be tough for an owner-operator to get out and acquire all of this information while also running a jobsite.

So you're not going to bid an in-town job the same as an out-of-town job.

If you're working at a large site, you're going to have room to move and roam around. You will have easier access to your materials, and get production fairly quickly. It is easier to maintain a certain amount of production per day or per hour.

When you get in a restricted area, you may have the same amount of material you have to move, or the same amount of pipe you have to put in - depending on what you're doing – but it's going to be much slower. You're going to be working with other contractors. You will have more limited access to trucks in and out, and more limited access to supplies in and out. Access to resources is going to be restricted because of dealing with traffic issues.

Say you're working on an out-of-town job, and you've got materials going off-site or on-site that you can move with fifteen or twenty trucks. Now if you're in-town, traffic is more of an issue, and it may take double the amount of trucks to get the same amount of materials in as it would ordinarily. You've got to account for that additional cost.

There are just a lot of restrictions with in-town jobs. Another thing to consider is your traffic control. When you're in-town, the local municipalities and government are going to make you have a certain amount of traffic control in place. That's going to cost money. You're going to have to submit traffic control plans. You're going to have to have flaggers and police officers to direct traffic, if that is an issue (and most of the time it is).

So if your biggest concern is having someone to help you coordinate the administrative issues, you can justify hiring one person to oversee that and your legwork for you. But if you're going to have to double the size of your company to do the job - take on a lot more manpower and more equipment – you should probably shy away from it. Taking on more than you can handle is a surefire way to go out of business or ruin your reputation.

One reason is that if you start renting, you're going to have to pay more than what you would if you owned the equipment, and that's going to reduce your income, reduce your profit, and increase your overhead. So you're losing money all the way around. If you do have more jobs in the queue, you should probably look into purchasing your equipment instead of leasing it. This will allow you to work at a better rate.

You're also going to increase your number of employees. The only way I would even consider this was if I knew that I was going to have other projects of the same size coming up, and I could justify that. You don't want to be in the position of hiring people only to have to lay them off six months down the road. They are depending on you to supply them and their families with a way of living, and you don't want to just have to turn your back on them, and say, "I'm sorry, I don't need you anymore." That's a lousy position to be in.

Again, don't bite off more than you can chew.

So the question becomes if it's a matter of taking a big job, out-of-town, or smaller job in-town, are you still better off with the smaller job, even if you have to hire someone to help you handle the administrative stuff?

Smaller jobs make good money, as long as you know what you're doing, and you use your time wisely. But regardless of job size, as a smaller or younger company, I would lean towards out-of-town jobs. When you get into town, it's more complex and you've got to have more people involved. Either way, just make sure your bids properly reflect the location and size of the job.

Try to grow at a steady, reasonable rate, but don't take on a job that's going to require you to do a whole lot of hiring, or get real big real fast, if you don't have a stream of work coming up after that. It's going to be a lot less headache on you, and make your business a lot more enjoyable.

Chapter 16 – Job Bonding

When you get into jobs of a larger scale, whether you're a prime contractor or a subcontractor, you're going to be required at some point to acquire a bond for that job. If you are a prime contractor, you're going to have to do a bid bond as well as a payment and performance bond.

A bid bond is a bond that allows you, as the awarded low bid, to back out of a contract. If you back out for some reason, the insurance company will cover the difference between your bid and the next lowest bid. It's security for the company you are working for if you decide that you are going to be unable to fulfill your part of the contract. Bid bonds are the responsibility of the prime contractor on a job. If you are a subcontractor, you don't have to worry about that.

As a prime contractor, if you are awarded a job and everything's good, you're going to be required to do a payment and performance bond. At this point, if you cannot perform any part of the job, or you are negligent on that project, then the payment and performance bond has to kick in to either compensate for the differences or to pay for another contractor to come in and do that work.

If you're hired as a subcontractor, you'll be required to do a bond portion. You are going to be required to carry your own payment and performance bond to account for your part of the work on the project. That gives the GC the same security that they're giving to the owner. At that point, it's the same thing – if you're negligent and cannot perform your work or abandon the job, then they have the authority to pull your bond and use your bond money as a resource to hire someone else to come in and complete your portion of the work.

For the most part, typically you will work with your own insurance company for bonding, but sometimes, depending on the size of your contract, the GC may ask you a bond rate. They may ask, "Can you give me a bond rate to add in for your portion of work?" Say your bond rate is 1.5%, and the bond rate you're working off of is a .95%. At that point, they may choose to incorporate you under their bond rate and force you to pay theirs. In the bidding process, as long as you've got your bond rate figured in at what it's going to cost you, then you're good to go.

If you're bidding as a prime contractor, bid requirements are going to be in your specs. They're going to be in your bid package. It's going to tell you that you've got to have a 5% bid bond and you've got to have a payment and performance bond. At that point you know that you have to get these quotes from your insurance company, or your bonding company, to be able to figure that in your price.

If you're bidding as a sub, it's your responsibility to ask to whom you're bidding and if you need to figure a bond rate. If you do not ask that question, then shame on you. It could be there. It could not be there. But it's always worth asking.

Say your portion is $3 million out of a $10 or $15 million project. If you've got a 1.2% bond rate, you're got to pay X amount of dollars per month for that bond. It can get pretty costly. You could be spending $60 - $100 thousand for your bond fees. If you don't include that in your bid, it's got to come from somewhere.

Your insurance company should be able to do your bonds. Most of your insurance companies know that if you are doing contractor style insurance, they go hand in hand. But do not take for granted that you can just go out and get a bond. You've got to provide financial reports. They've got to see your quarterly, yearly numbers. They've got to see everything that you do financially. That's going to determine your bond rate, or even if you're eligible for a bond.

These are things you need to take care of with your insurance company before you actually bid on a particular job. If you have not taken care of this, and you ask the question, "Do I need to incorporate a bond on this project?" and they say yes, then you need to opt out at that time because you have not been through the procedures with your bonding company to make sure you can do that.

Chapter 17 – When to Hire an Estimator

When first starting out, most small construction company owners will do their own estimating, in addition to the actual work on the jobsite. So the question then is when is the time right to hire a full-time estimator? In my opinion, you need to be in business for a while. Hiring an estimator is definitely not something I would suggest you do right from the start. You need to get your feet wet, and see what you can and can't do.

The way I look at it is if you do not know how to put a bid together on a project, and to pick up work, you probably shouldn't even be thinking about starting a business.

If you're able to go in and bid your own work, pick up some jobs, and create some revenue, then you have a much better handle on your entire business. So maybe a year goes by, and then you go back and look at what you've done in the past year and try to figure out if it is feasible to hire an estimator, if you need one, if having one would help you grow your business, etc. It's kind-of a tough situation sometimes to determine what that breaking point is, when you need to make a hire and when you don't. It's definitely going to take some financial research to figure out what you can afford and what you can't afford.

One can argue the case that hiring that estimator will generate enough additional business that that person will pay for himself. A safer approach is to make sure that you've gotten to the point where you have enough extra money that you can just hire that guy even if he doesn't generate any new business.

It's a gamble either way. I would definitely lean more towards trying to get your finances to the point where you know you can afford to make the hire. One good thing to look at is what an average base cost of an estimator is going to be. Here in our area, there is a vast range of salaries that estimators earn, but in general you're looking at probably somewhere between $70 and $80 thousand per year to hire a good estimator. On top of that, most are going to want a vehicle and fuel furnished, especially a project manager/estimator.

You've got to be able to pay your own salary, pay your workers that you have, pay all your bills – equipment payments, insurance, office expense, etc. Then you need to be able to put some back in your company. After you have accounted for all that, if you've got enough money left over to pay another salary, then I'd say it's a good time to start thinking about hiring an estimator.

A good estimator is going to pick up work for you. Your business is going to start growing if the estimator can do what he says he can. Once that happens, then you've got to make sure you are able to stay on top of your jobs, and that the financial stuff works out as close as possible to the way it was estimated. It bears repeating here – if you don't track your costs, you won't know if your estimates are good or not. So it is critical to use some kind of program like ProfitDig to track everything.

If you need an estimator bad enough, but your finances are iffy, then it may be a situation where you have to offer them a percentage share of your business. If you can't really afford to pay him his salary, then you may have to give him some other kind of incentive.

You might say something like, "Hey. I'm going to give you a percentage of our profits. We've got a $200,000 budget on this project with no profit overhead. This is just our budget. If you can bring that in under budget, I'll give you 20%." Or whatever that magic number might be for you if you come in under budget.

In this scenario, you've started out paying the estimator a small salary, with the potential to make much more based on performance. Put a time limit on it. Give your estimator X amount of time to prove himself and to help you build your company. If he can't do that, you might want to look at hiring another estimator.

Knowing how to properly estimate a job yourself will also help you in the process of hiring an estimator. Not all estimators are proven. The ones that are proven, with the construction industry the way it is today, already have jobs – probably good paying jobs. If you want to get one of those really experienced guys, you have to be willing to add more money to your jobs to account for the extra salary and overhead. Just starting out, that's going to be tough to do.

So you might be looking at a younger person who is just getting his feet wet in estimating. If you know how to do estimates yourself, then you can kind-of watch over him. Tell him you want to review his bids before they are ever seen. As owner, you should look at what he is sending in and make sure that you feel comfortable that you can make money with how he's got the project figured.

Assuming you don't have unlimited resources to just throw money at the situation, it can be tough to find an estimator. I would suggest talking to other general contractors. Tell them you are looking for a good estimator.

Put the word out. Word of mouth is always going to be best way to find someone. You've got material suppliers who deal with estimators everyday. I'd pick up some of my materials from buyers and say, "Hey. If you don't mind, I'm looking for estimators. Could you help put the word out?"

If a guy walks into your office and says he's an estimator, you need to be prepared to ask a lot of questions to evaluate how qualified you think he is. You can kind of throw out a few questions here and there about his work history, where he's been working recently. Find out what types of jobs he is accustomed to bidding and managing.

Find out the average size project he has experience with. Is it $5,000,000? Is it $1,000,000? Is it $200,000? You need to make sure that that estimator has at least worked on projects of about the average value of your jobs or greater. Tell him to give you some documentation. What jobs have you worked on in the last year or two? Who have they been for? You can tell by his lingo – how he talks about his experience – if he knows what he is talking about

If I had my choice, I would rather hire someone who has been hands-on and actually been in the field and done the work. If I'm estimating dirt work, it's going to impress me that an estimator has been out in the field. He's loaded trucks. He's cut grate. He's figured grate. He knows how long it's going to take to move this material from one side of the job to the other.

Someone who is fresh out of college who has never stepped foot on a job site is not going to know different terrains and the effects different geographic terrain can have on a project. He is not going to have a good feel for things like the need for extra trucks or extra manpower to move dirt certain distances.

It's a lot to process going from the field to the office, so I'd rather have someone who has had that field experience even if I've got to spend a little time to help them figure out dollar amounts and such. He's going to know how long it takes him to move dirt and actually perform the work.

Chapter 18 – Managing Your Finances

Let's talk a little bit about managing finances.

If you own a construction company, you've probably got a lot of money going out, and hopefully some money coming in, but at the end of the day, if you don't keep track of all that, you don't really know if you made any money or not. So how can you manage that?

There are several different ways. From my experience dealing with other companies as well as my own company, I can tell you that there are going to be times when get behind, when you're going to struggle. Maybe your pay request didn't come in when it was supposed to, and you only had a limited amount of resources there set aside to make payroll or buy fuel. Now you've had to use that up because your pay is running thirty, sixty days late.

Dave Ramsey is big in the world of personal finance, but he's a pretty smart man and some of his ideas can be applied to managing your company's finances. His envelope system can work for a construction company just like it does for personal use.

What I mean by that is when I start a new job, I make sure all accounts are accounted for, per bid item. I look at how much labor I've got to have, how much fuel I've got to have, how much equipment time I've got to have. If I need road trucks, I look at how many hours I've got to have for road trucks, and how much fuel I need for them. All of that is broken down per bid item.

It's tough to do, and it takes time, but it's worth it. Using a system like ProfitDig or something similar will greatly reduce the time required for this. If you don't have some kind of automated system, then you're going to have to sit down and do it on pen and paper or whatever works best for you.

But every month when you send your bill for a project, you've got to figure out how much work you have done on each bid item. Then at that point you see, "well, I'm 30% on this bid item," or "I'm 20% on this bid item." You've got to figure out how much fuel expense you have, how much equipment time, how much labor time. An easy way to do it is to set up separate checking accounts for each of these items.

There were times in my past when I have attempted to just keep up with all this information in a spreadsheet, but we kept neglecting to keep up with it properly. We figured at that point, let's just start extra checking accounts. It's going to cost a little bit of money for more checks and fees and such, but it's really nothing compared to what you can lose by not keeping up with your accounts.

So we set up a trucking checking account, a fuel account and equipment account. We also had our general funds, which also served as our payroll. Then we had a money market account in which profits, or anything left over, went into.

The way we did it is I would tell somebody in the office who was going to be writing checks, "You need to take out this much for trucking, this amount for materials, this amount for fuel, this amount for labor for this month's draw on this job." Then she would break it down by account and whatever was left over would go into the money market account. It was an interest bearing account and the money just sat there. If we needed it that was fine, if we didn't that was fine too.

After your initial first month or two, everything is going to offset itself. You're going to start seeing that money build up. Take your fuel, for instance. If fuel prices go down, that's going to help you because you estimated this job on a higher fuel cost, so that adds up. You estimated a certain amount of time for your equipment. If you run your jobsite efficiently, you can try to cut that time down. So at that point, you may be using less fuel, so your fuel account is starting to build up. You've got more fuel money, and you've got more equipment money. If you can cut your labor down, then the funds in your labor account will start to grow too.

Well, say six months down the road everything is looking good, then every time you need fuel, you have money sitting there in your fuel account, so you can pay for it no problem. If you have a flat tire on your dump truck, you have money sitting in that account to pay for it.

Everything breaks down. It's costly. Say you have a dozer and its' engine goes down. Well you're going to drop ten, twelve grand to replace that engine, maybe more. But then you look into your equipment account and see that, "Hey, I've got $40,000 sitting in my equipment account. I don't have to go to the bank. I ain't got to talk to no banker. I can just write a check for that."

It's very important to make sure you're money goes where it needs to go. That way your general funds, or your money market account, will start to build up. You should try to leave that alone. You want your salaries to come out of your labor account and your fuel costs to come out of your fuel account.

So every month, as money comes in, rather than just depositing that into a single checking account, you divvy that money out into the separate accounts. You need to make sure you put the correct amount into each account.

Say you're moving dirt, for instance. Say you got cut to shift. It all boils down to time because fuel is time, equipment is time, labor is time. If you're burning X amount of gallons of fuel per hour and you've got twelve hundred hours figured in this one bid item, then you know how much you're fuel is going to be. So whenever you bill for that month for a particular bid item, if you have billed 50% of it, then you're saying you are 50% complete with that bid item. You can look back and see what 50% of your fuel cost was, what 50% of your trucking cost was, 50% of your labor cost. At that point it all divvies out.

That's the amount that you're going to separate out of this check. This doesn't have to be a lifelong thing. This is to get you on track to show you what you need to do – how you need to run your business through what I call, like Dave Ramsey, the "envelope system." After a few months or a year, it gets to where it's habit. I'm not going to touch this money, I've got to pay for fuel, I've got to pay for this and that. Anything left over I'm going to stick over here and that way, it's there if I need it. That way everything is accounted for.

This system can work for any company. If you're already established, you may not need it. But say you are established and you're not tracking your costs and you don't know really what your costs are, what happens typically is your cost of living seems to always go up. You're drawing more money out for personal use and more toys. You say to yourself "Hey, I'm going to buy this piece of equipment or I'm going to buy that piece of equipment. I need a new truck." You see that money sitting there, so you find ways to spend it.

If you have a designated lane for your money to go – this money goes here, this money there, this money goes over here – then you are narrowed down. You get a smaller portion of that to spend. It's an eye opener. Then you may instead say to yourself "I may not need that new pickup. I may not need that new piece of equipment right now."

At that point, your company starts to grow. You start to get more resources. You're not having to go borrow money to buy fuel. You don't have to have a line of credit because you're always tight. You can do away with all that stuff as long as you have adequate money and you're tracking your job costs accurately and correctly. Then that eventually goes away. You could even do away with this system because at that point you're going to know you don't need to touch the bulk of your money.

Chapter 19 – Dangers to Avoid

One thing that brought about the idea for the ProfitDig online estimating and costing system was my losing my business. There are a lot of dangers involved in running a construction company. The business is not going to run itself. You've got to be aware of what's going on. You've got to know what clients you have, who you're dealing with and who you're comfortable with. You've got to know if your jobs are making money, and if not, why.

Putting All Your Eggs in One Basket

One thing that you've got to be aware of is something my Dad always warned about (even though we ended up falling into the same trap), which is putting all of your eggs in one basket. That was our downfall. We got in with one developer who had a lot of property and a lot of work. We quit bidding other work because he had enough work for us to do without bringing on any other jobs. We were running several crews and we were steadily buying equipment because we never had enough to do everything that we had going on.

In the downturn of the economy, the developer shut down. When he shut down, we shut down with him. We had all of our eggs in one basket. That was the only man we were working for, and we had no one else at that point.

In a down economy, there was not any work to be found. We were really struggling to find avenues to work in. For one thing, we had not had a job that we had to bond in six years' time, so we lost our bonding capacity. Municipal work was the only thing that was going that we had a shot at. But we couldn't do any of it because we weren't able to bond anything.

Even if things are really good right now, with a single customer who is sending you loads of work, you should really try to diversify and always be looking for other customers to do work for.

2007 and 2008 were our most profitable years. Then around 2009 - 2010, it was like someone cut a switch off. We never saw it coming. One day we had plenty of work and the next day we were sitting on our butts with nothing to do.

You can't predict the market and what it's going to do, so you need to have other avenues. You need to be working for different clients, at least two or three or however many you possibly can. You don't want to be stuck in one hen house with nowhere to go.

Slow Pay Customers

Slow pay customers are another potential danger. If a customer is slow to pay, there's some reason that's causing that and most of the time it is because he's not managing his business like he should. You don't want to end up in a lawsuit if you can avoid it. It's going to cost you money. The best thing to do is try to pick up other work and cut your ties with that client because there's a pretty good chance it's not going to change. If you've done two or three jobs for him and it's always a hassle to get paid, it's pretty much going to continue that way.

Again, it comes down to diversifying so that you're not too reliant on one or just a very small number of customers.

Your Company's Image

You want to be well known. You want to be respected. But it takes time. As a project manager or construction company owner, when you ride up and down the road, you can't help but notice other projects similar to the kind of work you do, and you look at those. You look at who's doing the work, what their equipment looks like. Do they have manpower that's working or do you see men constantly standing around? At that point, that suggests to me that if they've got good looking equipment and their men are constantly working, this is a good company. I may want to try to work with them and do business with them.

It's been my experience that whenever we've had new equipment or new trucks on the job site, I've actually had developers stop by and hand out cards and introduce themselves and say, "We'd like to send you some drawings and get you to bid on some work for us."

Just presenting yourself well on the job site, looking professional, with everybody busy doing their part, can actually earn you more work.

Similarly, if you don't present yourself well, that potentially costs you work. When you are first starting out, you may not have the best equipment. But if you're doing quality work, you're going to pretty quickly get to a point where you can afford to do better. When you have men on the job that are standing around, that act like they are lost, not 100% sure what's going on, your equipment is not in the best shape, ragged or always broke down, and you've constantly got a mechanic on site working on something, then you're losing production.

You're costing yourself money. You're costing that developer or property owner or whoever you're working for money because they have a deadline. They're like everybody else. They've got money sitting in an escrow account from which they draw funds, so they have interest to pay. The longer that job drags on, the more interest they've got to pay and the more money it costs them. That is not the way to build an image for your business.

You want to do quality work. You don't have to buy new equipment. There is a lot of good used equipment in the marketplace, and if you can't afford to buy then rent equipment. Do whatever you have to do to represent your company in the best way possible to get repeat customers.

Employee Risk

There is also a risk in hiring bad personnel. The people you put in play on your crews, as the old saying goes, can make you or break you. If they're company people and they want to work and work for you, they're going to do a good job for you.

You need to make sure that the people you hire are qualified. If you need an operator, don't go out and hire a brick mason because then he's got to learn how to run a machine. Or say it's a pipe layer that you need to hire. A brick mason isn't going to know a thing about laying pipe. You need to make sure that the people you hire are qualified for the work you've got them doing.

You've got to make sure that you have adequate manpower to do the job you bid in the amount of time budgeted. You must meet your deadlines. You've got X number of days or X amount of time to do this project and if you don't have enough manpower or have enough equipment, you're hurting yourself. You're costing yourself money. Again, you're also costing the developer or the owner money. You've got to stop it quick. You're bleeding and you've got to stop that bleeding before you bleed out.

You risk ruining the relationship that got you the job. If you don't have the resources to do a project, then the person who hired you is going to get a bad taste in his mouth. If you really want to work for this particular person again, it's going to be more difficult. His thinking may be along the lines of,

"Well, I hired Jeff Spencer to come in here and do this project and he pushed me 30 days late or he pushed me 60 days late and I don't think I want to take that risk with him again." If you don't have the manpower, you've got to find somewhere to get it and get it quick because every day that you prolong a job, that's time and money wasted.

So...the next question is where to find manpower. If all you need is laborers, then you will be just fine going through a temp agency. However, my experience with temp agencies with regard to specialty employees has not been favorable. When I bring anyone new on the job, I try to give him the benefit of the doubt. I work him 2 or 3 days. If he can't do the job that he is supposed to be qualified for, then he is gone.

Think about it – as of the time we are writing this, construction is booming (at least where we are). There is a shortage of workers. A skilled, qualified construction worker does not have to go through an agency. If you've got labor work or what I like to call "busy work", then you can pretty much find somebody to work for 9 or 10 dollars an hour and clean the job site up or pick up trash. That's not the problem. So for that reason, keep your eyes open for skilled men you can hire along the way. Or make relationships with those guys. Keep in touch with the guys you used to work for.

Equipment Risk

There is also risk in buying equipment. If it's a job that's going to be long term, say 1 to 2 years, or it's a big project, you might be safe to make that initial purchase. But if you don't see your company growing or keeping that work load, or you're not sure if you will be able to pick up another job to move that equipment to when that job's done, I would recommend renting.

If there's a question in mind, most all of your equipment companies will do a lease purchase. That's the way I set most of mine up. I'd lease it with an option to buy. Most of them will work a deal with you. You lease for 9 months and at 9 months you either turn it in or buy it. Most of them will let you take 90% to 100% of what you paid on that lease towards a down payment on a machine.

That's a way to cover yourself because you can lease the equipment to cover the work you have and if it looks like this is going to turn into more of a long-term relationship, then you can go ahead and buy it. Or if the job ends and you don't need it any more, you can just turn it back in.

Another good thing about a lease conversion is that when you roll that over and finance it, your payment usually goes down somewhere from 15% to 25% to purchase versus what it was costing you to rent.

Learning From My Mistakes

ProfitDig partner Jerry Work interviews Jeff Spencer about his background and experience, and what led to the creation of ProfitDig.

Jerry: Jeff, I thought our readers might be interested in learning just a little bit about your experience. At one point, you were co-owner in a construction business with your father and brother, is that right?

Jeff: Yes.

Jerry: Tell me what went down there.

Jeff: We were a general contractor. We did excavation work. We did municipal work, utilities, and a little bit of structural stuff - not a whole lot, but most of our structural stuff we would sub out. We really took off. My father and his former business partner were partners for right at twenty-eight years and were very successful. They were getting close to retirement age and decided to disperse and go their own way. My Dad and my brother and I partnered up, and from the get go, through his resources - which we've talked about how important resources are - we were able to pick up work pretty quick. Of course, we were a lot smaller than what they (my father's former company) were, but within a couple of years, we grew and got pretty close to the size that he was originally.

We were trying to keep our profit up as much as possible and keep overhead as low as possible, so we were pretty much all working out in the field. My wife was kind-of doing the books, but nobody was really tracking job costs. We were still bidding work day in and day out.

I'd literally bid out of the back seat of my truck. I'd bid on a job, get a few free minutes, drag out a set of plans and look at them, and try to throw some numbers together. We were working all over our area. We had three to four crews of our own and had three or four crews subbed out doing work for us. There again, we were steadily working, had money steadily coming in, but we weren't tracking any of it.

After about six years of that, it caught up to us. We got to where we were struggling paying our bills. We never could get enough money to pay for our materials when we needed them. It was a struggle to get payroll. Payroll taxes were killing us. We were just using that money that we should have been putting aside to cover our costs.

So what happened to cause this change of circumstances? When we first started off, we were where we needed to be. But as time went on, inflation occurred and prices went up on materials, but we were still bidding the same that we were three or four years previous. Another problem was that we had been giving some of our laborers and operators and pipe layers, people who had been with us for a while, cost of living raises because we felt they were deserved. But again, we weren't tracking our costs, and we did not know what effect that was having on us.

To make a long story short, after eight years, it put us out of business. We went from thirty-plus employees down to a handful of employees down to nothing. We were out of business and didn't know what happened. We had no explanation.

Jerry: If you had it to do over, what would you have done differently?

Jeff: If I had to, I would have hired somebody just to take care of our job costing. I looked at some of the programs out there to do job costing on. I'm not going to mention any names of the ones I looked at, but $10,000 was about the cheapest program I could find, and by the point when we knew we needed it, $10,000 was a big commitment. We did not have that kind of capital at that point to do it. We were too far gone when we realized that that's what we needed. If I could go back and do it over, the very first thing I would do is put in my budget some kind of estimating and job costing program to make sure that I'm able to make the money I need to be making and not losing money.

The point for anyone reading this is that if you're bidding on jobs, you have got to track your costs so that you have clarity on what's going on - if you're making money, if you're losing money. If you're not making money, you won't last long.

Jerry: I guess then, Jeff that is sort of the inspiration behind Profit Dig.

Jeff: Yes, it is.

Because it is a cost effective way for anyone, especially a small construction owner, to bid jobs and track his costs against those jobs, and it doesn't cost thousands of dollars.

In my old business, we needed help tracking job costs and creating estimates, but we were too far gone. We could not afford to make those purchases, but Profit Dig was built and designed to do that for you at an affordable rate. Even if you are a single user - the guy who owns the company, but who is also out in the field running the company - you can afford this system. It's not an up-front lump sum cost. It's just a small monthly fee.

Jerry: All right, Jeff. I appreciate your sharing your story with us, and I think you've made your point. I hope everyone reading this understands the importance of what we're talking about here, or they'll suffer the same results.

Jeff: Most definitely.

About the Authors

Jeff Spencer

Jeff Spencer is a professional construction estimator with over 25 years of experience in the construction industry, and a founding partner in ProfitDig LLC. In the early 2000's, Jeff, his father Benny, and brother Dowell built a successful construction company and then lost it all when the economy took a downturn. From those hard-learned lessons came the idea for ProfitDig, an online system to make it easy and affordable for construction companies of any size to bid jobs and accurately track their costs.

As a young man, Jeff was a national fiddle and buck-dancing champion, raised cattle and went on to acquire three national champion bull titles in the Red Poll cattle breed. He has a wife, Sarah, and three children, and currently lives in the small town of White Bluff, Tennessee.

Jerry Work

Jerry Work is a professional Internet marketer and developer, and a partner in ProfitDig LLC. He enjoys playing guitar and collecting interesting vintage items. He has a wife, Wendi, and two children, and lives on a farm in Dickson, Tennessee.

SPECIAL OFFER

The topics I have discussed in this book are so important for running a successful construction company that I want to extend you a special offer to try ProfitDig. As I have mentioned a few times, ProfitDig makes it easy to create professional bids and track your job costs. If I had had this tool when I was running my company with my Dad and brother, we would still be in business today.

Try ProfitDig today and get it for half price for the first six months. If you actually take the time to use the system, you will never want to be without it. ProfitDig is a no contract monthly service, so it will hardly make a dent in your budget. But it will make a HUGE impact on your bottom line.

Visit **www.ProfitDig.com/bookoffer** to take advantage of this offer.

Even if you don't use ProfitDig, I implore you to use SOMETHING to manage your bids and track your costs. The success of your company depends on it.

Good luck!

www.ingramcontent.com/pod-product-compliance
Lightning Source LLC
Chambersburg PA
CBHW070327190526
45169CB00005B/1775